Traceability in food industry

Preetha Palanisamy
Deepa J.

Traceability in food industry

LAP LAMBERT Academic Publishing

Impressum / Imprint
Bibliografische Information der Deutschen Nationalbibliothek: Die Deutsche Nationalbibliothek verzeichnet diese Publikation in der Deutschen Nationalbibliografie; detaillierte bibliografische Daten sind im Internet über http://dnb.d-nb.de abrufbar.
Alle in diesem Buch genannten Marken und Produktnamen unterliegen warenzeichen-, marken- oder patentrechtlichem Schutz bzw. sind Warenzeichen oder eingetragene Warenzeichen der jeweiligen Inhaber. Die Wiedergabe von Marken, Produktnamen, Gebrauchsnamen, Handelsnamen, Warenbezeichnungen u.s.w. in diesem Werk berechtigt auch ohne besondere Kennzeichnung nicht zu der Annahme, dass solche Namen im Sinne der Warenzeichen- und Markenschutzgesetzgebung als frei zu betrachten wären und daher von jedermann benutzt werden dürften.

Bibliographic information published by the Deutsche Nationalbibliothek: The Deutsche Nationalbibliothek lists this publication in the Deutsche Nationalbibliografie; detailed bibliographic data are available in the Internet at http://dnb.d-nb.de.
Any brand names and product names mentioned in this book are subject to trademark, brand or patent protection and are trademarks or registered trademarks of their respective holders. The use of brand names, product names, common names, trade names, product descriptions etc. even without a particular marking in this works is in no way to be construed to mean that such names may be regarded as unrestricted in respect of trademark and brand protection legislation and could thus be used by anyone.

Coverbild / Cover image: www.ingimage.com

Verlag / Publisher:
LAP LAMBERT Academic Publishing
ist ein Imprint der / is a trademark of
OmniScriptum GmbH & Co. KG
Heinrich-Böcking-Str. 6-8, 66121 Saarbrücken, Deutschland / Germany
Email: info@lap-publishing.com

Herstellung: siehe letzte Seite /
Printed at: see last page
ISBN: 978-3-659-49883-1

Copyright © 2013 OmniScriptum GmbH & Co. KG
Alle Rechte vorbehalten. / All rights reserved. Saarbrücken 2013

Traceability in Food Industry

By

P. Preetha,
Research scholar,
Department of Food and Agricultural Process Engineering,
Agricultural Engineering College and Research Institute,
Coimbatore- 641 003.

&

J.Deepa ,
Research Scholar,
Department of Food and Agricultural Process Engineering,
Agricultural Engineering College and Research Institute,
Coimbatore- 641 003.

Contents

Sl.NO	TITLE	Page
1	Introduction	1
2	History of traceability system	3
3	Related Laws and other rules	5
3	Traceability drives in Food industry	13
4	Characteristics of the Traceability system	15
5	Food Traceability Frame work	21
6	Traceability System and its types	22
7	Traceability Process steps	27
8	Internal Traceability in Food Manufacturing	31
9	Linking Identification and information	35
10	Within a business- process traceability	36
11	Traceability in Mineral water Plant	38
12	Traceability in Chocolate industry	42
13	Examples of food traceability tools and Labels	47
14	Conclusion	47
15	Reference	48

1. Introduction

"Traceability" is to "trace", which in figurative sense is a "mark left by an event." Tracer is also "indicate the way forward" or "mark the contours. The word "traceability" is relatively recent in the French language, and is since 1998 in the Petit Robert, which defines it as "the possibility of identifying the origin and to reconstruct the route (a product), since its production to its dissemination."

ISO 9000: 2000 defines traceability as "the ability to find the history, implementation or location of what is considered". It is therefore a process which includes the opportunity to trace the different stages of life and places a product from its creation until its destruction. In other words, traceability can identify, for a product:

- ✓ All stages of its manufacture,
- ✓ The origin of its components and their suppliers,
- ✓ Where the product and its components have been stored,
- ✓ Checks and tests on the product and its components,
- ✓ Equipment used in manufacturing or handling,
- ✓ Direct customers who bought the product.

Traceability is a concept that can be applied to all sectors: chemical, pharmaceutical, automobile, etc..We will focus particularly on traceability in the food industry, which is currently component of ensuring food safety. In the food chain, traceability means the ability to trace and follow a food, feed, food producing animal or substance through all stages of production and distribution. Stages of production and distribution means any stage including import, from and including the primary production of food, up to and including its sale or supply to the final consumer and, where relevant to food safety, the production, manufacture and distribution of feed. Many manufacturing systems, including food manufacturing, have sought registration to the ISO 9001 Quality Standards. These require that the product should be able to be traced from the current stage back through all its stages of manufacture through accurate and timely record-keeping. The requirement for paper documentation has recently been changed; computer records solely can now be used as evidence of compliance. In primary production, traceability has been defined as the ability to trace the history of the product through the supply chain to or from the place and time of production, including the identification of the inputs used and production operations undertaken (British Standards Institute PAS 85:2000). Legislation has been recently introduced to ensure livestock identification and the tracking of livestock movements. Many of the farm assurance schemes require some level of traceability to be in place within primary production.

Traceability systems are part of systems which enable industry:
- ✓ To comply with relevant legislation
- ✓ To be able to take prompt action to remove products from sale and protect brand reputation (through a failure in product quality or food safety incident)
- ✓ To minimize the size of any withdrawal and hence the costs incurred in recovering, disposing or reconditioning products already placed on the market.
- ✓ To diagnose problems in production and pass on liability where relevant.
- ✓ To create identity preserved non-GM sources of soya and other ingredients.
- ✓ To minimize the spread of any contagious disease amongst livestock.
- ✓ To protect the food chain against the effects of animal disease.
- ✓ To assure meat and meat products and maintain markets and consumer confidence.

To create differentiated products in the market place because of the way they have been produced. The implementation/upgrading of traceability systems within the industry may occur with changes in process control systems (as the size of the processing operation outgrows the current manual systems). Customers may require the delivery of goods into warehouses operating with standard barcodes *etc.* hence requiring the purchase of bar code printing and reading equipment.

2. History of traceability:

In 2005, The European commission implemented several directives and regulations on traceability of foodstuffs. In addition to legal authorities, consumers and Non Governmental Organizations (NGO), retailers also demand that food producers must have systems for record keeping of process related parameters and traceability functionality in place.

Information frequently asked for are i.e.; source of raw materials, additives used, time/temperature log, production date and time, use of Genetically Modified Organisms (GMO), slaughtering and other processing conditions, animal welfare related issues, other ethical aspects like use of child labor and environmental issues. Such systematic record keeping requires a new behavior, which is a big challenge for all parties in the whole supply chain.

Food safety related issues, in particular several serious food scandals, started the traceability focus. One can list up; dioxin in Belgian chicken; E.coli 0157:H7 in Hudson Foods Company beef (USA), Mad cow disease, Foot and mouth disease, Scrapie, Asian bird influenza, and numerous occurrences of pathogenic bacteria like salmonella and listeria. In all these cases negative consequences could have been reduced if appropriate traceability systems were in place, and isolation of the contaminated batches made possible. It is also worth noting that "innocent" companies often get involved in food scandals. This is a key concept in traceability; if you are not able to document that your products are "clean", you might be considered as "guilty" and forced to carry out expensive recalls.

In the Belgian dioxin case in 1999, motor oil containing 1 g of dioxin came into a recycling plant for vegetable oil by mistake. The fat produced went into chicken feed production. Effects were not severe, but the scope was enormous, and at least 1600 chicken farms were contaminated. After digging into the case, legal authorities enforced withdrawal of all feed batches, chicken or eggs of Belgian origin. Since the industry had poor routines for linking of raw materials and ingredients to production batches, even "innocent" companies with insufficient record keeping had to recall their products. The result of this case was that the whole Belgian poultry industry had to close down - to a cost of approximately 1.3 billion USD.

Additionally, supermarket chains presently push requirements on traceability down the food chain. The influential and substantial retailer initiatives EurepGAP and Global Food Safety Initiative (GFSI) have really put ethical food production and traceability on the agenda. Food producers now experience more strict demands related to food safety, ethical production and documentation (record keeping) of origin and process related information from supermarkets than from legal authorities. The reason for this is that supermarkets have to deal with constantly more demanding customers and even more importantly; every day they have to carry out costly and time consuming withdrawals of contaminated food-stuffs.

Many will claim that food safety is the most important trigger for implementing traceability systems. However, with new business practice, where actors in the supply chain are being more and more integrated, i.e. they share internal production data, the access to traceability information recorded earlier in the chain is definitively considered to be a competitive advantage. A recall is defined as a publicly known, often publicly ordered removal of a product from the market, where as a withdrawal does a product removal initiated by the food business itself, often without anyone outside the company know about it. Both are costly for the company, and if a recall happens, the company image and brand image suffer as well. If traceability data is easily accessible, preferably on a standardized electronic format, the number and scope of recalls and withdrawals can be reduced, and in some cases, withdrawals can be done without public notice.

There is a clear trend towards supermarkets and other types of large buyers requiring that suppliers must have systems for electronic exchange of trade information. For these businesses, handling an enormous number of products and transactions, electronic ordering and payment as well as the ability to re-use data, is a major cost saver compared to manual practice.

Improved data recording, keyed to traceable units, gives us new and more meaningful data material, and can be used in several applications:

- Benchmarking of suppliers
- Benchmarking of own production facilities
- Production optimization
- Purchase, warehousing, logistics and sales

On the highest level, food businesses working seriously with traceability, documentation of ethical or quality oriented production practice and profiling of such, may over time benefit from this when building reputation and brands.

3. Related Laws and other rules

The laws concerning traceability systems are as follows:

The Law Concerning Standardization and Proper Labeling of Agricultural and
i Forestry Products (JAS Law)

In this law, standards for the agricultural and forestry products as well as the standards regarding proper labeling on the quality of the agricultural and forestry products are laid down. It is obligatory to have foods and beverages sold directly to consumers labeled in accordance with the Quality Labeling Standard under the JAS Law. Specific contents are laid down in "The Quality Labeling Standard for Fresh Foods" and "The Quality Labeling Standard for Processed Foods" based on this law, and indication of the name, origin, and so forth (some processed foods are required to label the origin of the ingredients) is required. In this law, in case of the disclosure of an incident such as false labeling of origin, prompt revelation of the name of the involved food business operator and the penalties, are laid down.

ii Agricultural Products Inspection Law

In this law, a system of "agricultural products inspection" (grade inspection and constituent inspection) is established in order to contribute to fair and smooth trade and to improve the quality of agricultural products such as rice. According to "The Quality Labeling Standard of Brown Rice and Polished Rice", unless the brown rice is certified according to this law, place of origin, variety and harvested year shall not be labeled.

iii Agricultural Chemicals Regulation Law

In this law, registration system for agricultural chemicals and regulations on sales and usage of agricultural chemicals are established. A person who uses agricultural chemicals shall not use them against the regulation determined by ministerial ordinances. The Minister of Agriculture, Forestry and Fisheries or the Minister of Environment has a right to require any person using agricultural chemicals to submit

- When exporting, it is necessary to consider the laws and ordinances of the country and regions concerned.
- a report on the use of the agricultural chemicals, or to have the necessary materials such as agricultural chemicals and ledgers inspected.

iv Fertilizer Control Law

In this law, standards, registration, restrictions on input/application, and labeling standards of fertilizers are established. A person who uses fertilizers (such as producers) is prohibited from using "specific ordinary fertilizers" without guarantee labels. Specified ordinary fertilizers are designated by a government ordinance, to contain ingredient(s) that will remain as residues and may be harmful to human and animals). The Minister of Agriculture, Forestry and Fisheries or the governors of prefectures may require any person using fertilizer to submit a report, or to have the premises inspected when the necessity is recognized in accomplishing purposes of this law.

v Pharmaceutical Affairs Law

In this law, regulations regarding manufacturing, importing, sales and proper use of animal medicines are established. In the ministerial ordinance concerning the regulation of the usage of animal medicines, which is based on the law, the prescribed usage, dosage, withdrawal period and so on are determined in the ministerial ordinance requires an effort to record of the following items in the ledger when medicines are used.

- Date medicine is used
- Location medicine is used
- Kind, number and other distinction of the animals on which medicine is used
- Name of medicine
- Usage and dosage of medicine
- Date to be slaughtered, to be unloaded or to be shipped in order to be served as food

vi Law Concerning Safety Assurance and Quality Improvement of Feed (Feed Safety Law)

This law prohibits the compounding of antibacterial products in animal feed, regulates feed additives, and establishes standards of toxic substances. According to the revised ministerial ordinance concerning ingredient standards of feeds and feed additives issued in 2003, anyone who uses feeds must make an effort to record the following and keep the record.

- Date when feeds are used
- Location where feeds are used
- Kinds of livestock feeds are given
- Name of feeds
- Amount of feeds given
- Date which feeds are received and name of the person or organization from which feeds were provided

vii Slaughterhouse Law (Abattoir Law)

This law determines regulations regarding the establishment of abattoirs, sanitation management in abattoirs, sanitation management of slaughter or dissection of livestock, and inspection of slaughter or dissection of livestock. The governors of prefectures may, to the extent necessary for the enforcement of this law, collect necessary reports from owners, managers, slaughterers or other parties concerned. Also they may have officials concerned to inspect facilities, ledgers, documents and other objects.

viii Law on Special Measures against Bovine Spongiform Encephalopathy

To prevent the occurrence and spread of Bovine Spongiform Encephalopathy, this law establishes special measures such as prohibition of feed containing cattle meat and bone meal as well as regulations for report and inspections of dead cattle and inspections for BSE at abattoirs.

This law establishes that cattle owners or managers (in cases that cattle are managed by entities other than the owners) shall ensure that each of their cattle wears identification ear tags, shall record the specific information (Date of birth, moving record and other information) and shall provide information necessary for managing.

ix The Law for Special Measures concerning the Management and Relay of Information for Individual Identification of Cattle

In this law, identification of cattle and beef, proper management and transmission of information are determined. A person who manages cattle is required to notify the Minister of Agriculture, Forestry and Fisheries regarding its birth, import, transfers or receipts. Also slaughterers, sellers and suppliers of specific cuisine are required to indicate Individual Identification Number (or corresponding lot numbers) to the beef they handle, and to record the items stipulated by the government ordinance concerning transfer and selling (i.e. Individual Identification Numbers, date of transfer, name of purchaser, weight of beef and so on), and to record and store the record.

x Poultry Meat Inspection Law

This law determines permission of poultry handling and processing, observance matters for processing managers such as sanitation, poultry inspection and so on. Governors of prefectures may, to the extent necessary for the enforcement of this law, require processing managers to submit a report on their business situation or have officials concerned to inspect (inspections of facilities, ledgers, documents and other objects).

xi Food Sanitation Law

This law establishes necessary regulations and any other necessary measures in order to secure food safety from the viewpoint of public health. Based on the regulation, "the guideline on making and maintaining records by food business operators" is determined. The guideline indicates the items about which food business operators are generally required to record, as well as the number of years such records must be kept.

Xii Health Promotion Law

In this law, the fundamentals on comprehensive promotion of the citizens' health are defined. Regarding nutritional displays, such as nutritional ingredients, which are attached to food products, the law establishes mandatory criteria such as items and methods to be displayed.

Xiii Act against Unjustifiable Premiums and Misleading Representations (Premiums and Representations Act)

This law establishes regulations and prohibitions on unjustifiable premiums and misleading representations in connection with transactions of a commodity or 10 services. This law prohibits any of the following indications: showing that the content of the product is remarkably better than that of the actual product in question, indicating that the product is far better than that of the competitors which is contrary to the fact. The law authorizes Japan Fair Trade Commission to require food business operators to submit materials which show reasonable evidence if that the commission needs to decide whether the claim is unfair or not. If the food business operator does not submit the materials, an exclusion order can be applied to the operator.

2-2 Standards, guidelines, etc. on food traceability systems

Voluntary standards or guidelines concerning food traceability systems are as follows;

(1) National standards

i "Requirements for Food Traceability Systems"

This document is the standard for verification of food traceability systems. It was made by the committee on the Third Party Certification of Food Traceability

Systems and was released in October of 2006. The standards can be used for self verification (check by operators themselves), for verification by business partners (done by direct relevant people) and for the third party verification. In case of the third party verification, if all the requirements are met, then the operator could be basically considered as having the food traceability system in place.

(2) Guidelines by items and by stages

i Handbook for Introduction of Traceability of Domestic Beef

ii Guidelines for Information Tracing-back System of Receipt, Shipment and History of ingredients

iii Guidelines for Food-Service Industry towards Food Traceability Construction

iv Guidelines for Introduction of Food Traceability of Fruits and Vegetables

v Guidelines for Traceability of Shellfish (Oyster, Scallop)

vi Guidelines for Introduction of Food Traceability of Egg

vii Guidelines for Traceability Systems of Farmed Fish

viii Guidelines for Introduction of Food Traceability Systems of Laver

(3) International Standards and Rules

i Codex Alimentarius Committee "Principles for Traceability/Product Tracing as a Tool within a Food Inspection and Certification System"

ii ISO/DIS 22005 Traceability in the Feed and Food Chain—General Principles and Basic Requirements for System Design and Implementation 2-3 Standards related to food traceability

(1) National Standards

i Specific JAS with the Disclosed Production Information

This is a standard regarding accurate transmission of food production information (producer name, the place of production, input/application information on agricultural chemicals and fertilizers) transmitted voluntarily by a food business operator to consumers. The Ministry of Agriculture, Forestry and Fisheries

(MAFF) registered certifying bodies (The third-party organizations) conduct certification. As of March 2007, there are three standards of this kind; beef, pork and agricultural products (whole fresh agricultural products such as rice, vegetables, fruits, mushrooms, etc.).

(2) International Standards and Guidelines

i ISO 9001:2000 (JIS Q 9001:2000)

This is an International Standard model for quality management and quality assurance determined by ISO (International Organization for Standardization). Ensuring traceability could be added as one of the requirements.

ii ISO 22000:2005

This is a standard for food safety management systems. The analyzing methods of food hazards are introduced from HACCP, the principles of which were determined by the Codex Alimentarious Committee. And the approach of the management system is introduced from ISO 9001. 7.9 is the requirements of traceability systems.

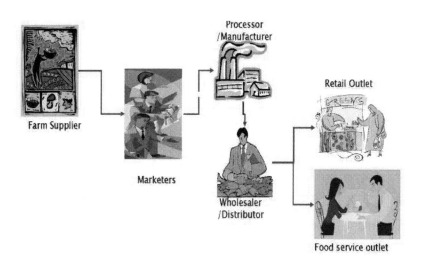

4. Traceability Drives in Food Industry:

The main drives includes

(i) Food Safety

(ii) Regulatory Compliance

(iii) Efficient Recall Management

(iv) High Customer Expectations

(v) Retail Mandates

(i) Food Safety

Concerns over safety of food consumed are one of the key drivers for traceability. This has gained importance with increased globalization leading to production shifting to low cost nations across the globe. Traceability has been mandated as a fall back measure to prevent contaminated food being consumed by executing a quick product recall. "Prevention is better than Cure" –enhanced measures to track food safety can avoid the situations leading to a product recall & thus the complex task of identifying batches to be recalled.

(ii) Regulatory compliance

One of the major drivers towards implementing traceability systems is Regulatory compliance. Sensing the risk involved in food contamination, governments across the globe have (or in the process of having) strict laws that mandate Traceability. EU General Food law 178/2002 US bio terrorism Act laws USFDA Country Of Origin Labeling (COOL) Food Allergen Labeling and Consumer Protection Act (FALCPA)

(iii) Efficient Recall Management

In the unfortunate event of a recall, companies who have sophisticated traceability systems would be able to exactly locate the batches that could possibly contain the contaminated ingredient. Only those batches can be recalled thus minimizing the cost & inconvenience of recalls. Assuming a company that doesn't have a robust traceability in place, would end up recalling everything from the retailers. If the contamination is reported in one of the key basic ingredients that go into multiple products, then the manufacturer doesn't have an option but to recall all product ranges containing that ingredient. Financial implications of such a move are severe on the companies. Recalls also result in significant erosion of brand value due to the reduced confidence among the consumers.

(iv) High Customer Expectations

In today's competitive world, customer expectations are soaring to new heights. In matters regarding food, customers are more demanding and would want manufacturers to guarantee food safety and at no additional cost. Quick recall through traceability systems would help in restoring customer confidence in the event of an outbreak. Though food companies have designed strict quality control measures to prevent contamination, they would need a fall back measure to ensure quick response in case of an emergency. In short, high customer expectations act as a key driver for companies to adopt traceability as a fall back for product recalls.

(v) Retail mandates

Gone are the days when products in the market used to be supplier driven – this is a demand driven era. Many Retailers driving the supply from the manufacturers and determine how products should be delivered at their warehouses /stores. Retailers are of the view that identifying items by RFID tagging would be the first step towards traceability. Apart from enhancing traceability, RFID tags also helps retailers by facilitating automatic checkout & enhanced theft control.

Traceability can be considered in four distinct contexts and in each it has a slightly different application:

✓ *For products:* it creates a link between materials, their origin and processing, distribution and location after delivery.

✓ *For data*: it relates the calculations and data generated through a quality loop and may link these back to the requirements for quality.

✓ *In calibration:* it relates measuring equipment to national, international or primary standards, to basic physical constants or properties or to reference materials.

✓ *In IT and programming*: it relates design and implementation processes back to the requirements for a system.

Objectives

Traceability system should be able to achieve the objectives from a technical and economic point of view. Traceability system should be verifiable, applied consistency and equitably. It should be result oriented, cost effective, practical to apply and complaint with any applicable regulation or policy and with defined accuracy requirements. Like ISO standards, traceability system should be established in order to achieve some specific objectives.

Objective can be set in a way which can support food safety and quality objectives directly or indirectly, should meet customer satisfaction, improve the effectiveness, productivity and profitability of the organization, and facilitate the withdrawl and recall of the product. Objectives can also be to identify the responsible organizations in the feed and the food chain, meet the national and international regulations facilitate the verification of the specific information about the product and communicate information to the stakeholders and consumers.

5. Characteristics of traceability systems

The basic characteristics of traceability systems, *i.e.* identification, information and the links between, are common in all systems independent of the type of product, production and control system that are served. In practice, traceability systems are record keeping procedures that show the path of a particular product or ingredient from supplier(s) into the business, through all the intermediate steps which process and combine ingredients into new products and through the supply chain to consumers.

The traceability of products is based on the ability to identify them uniquely at any point in the supply chain. The manufacturer or importer determines the size of a batch, which is identified uniquely. Throughout the food chain, new identities are constantly created as ingredients are combined in recipes, goods are bulked up for delivery, and/or large batches split to a number of destinations. Traceability requires both that the batch can be identified and that this identification gives a link to the product history.

Both products and processes may form key components (known technically as core entities) in a traceability system with information stored in relation to each. In the simplest systems, the only information carried is that showing the linked path along which products can be identified through the chain of manufacture, distribution and retail (*i.e.* information on the identity of the components, where they have been and when). Additional information may be carried *e.g.* information enabling processing efficiencies to be calculated for manufacturing systems, information concerning ingredient quality or origin. The amount and type of information can be extended as required by the system, and it may be carried for only part of, or throughout the whole, food chain.

A system for tracking every input and process to satisfy every objective would be enormous and very costly. Consequently, firms across the U.S. food supply system have developed varying amounts and kinds of traceability. Firms ydetermine the necessary *breadth, depth,* and *precision* of their traceability systems depending on characteristics of their production process and their traceability objectives.

Breadth describes the amount of information collected. A recordkeeping system cataloging all of a food's attributes would be enormous, unnecessary, and expensive. Take, for example, a cup of coffee. The beans could come from any number of countries; be grown with numerous pesticides or just a few; be grown on huge corporate organic farms or small family-run conventional farms; be harvested by children or by machines; be stored in hygienic or pest-infested facilities; and be decaffeinated using a chemical solvent or hot water. Few, if any, producers or consumers would be interested in all this information. The breadth of most traceability systems would exclude some of these attributes.

Depth is how far back or forward the system tracks the relevant information. For example, a traceability system for decaffeinated coffee would extend back only to the processing stage. A traceability system for fair-trade coffee would extend only to information on price and terms of trade between coffee growers and processors. A traceability system for fair wages would extend to harvest; for shade grown, to cultivation; and for nongenetically engineered, to the bean or seed. For food safety, the depth of the traceability system depends on where hazards and remedies can enter the food production chain. For some health hazards, such as Bovine Spongiform Encephalopathy (BSE, or mad cow disease), ensuring food safety requires establishing safety measures at the farm. For other health hazards, such as foodborne pathogens, firms may need to establish a number of critical control points along the entire production and distribution chain

Precision reflects the degree of assurance with which the tracing system can pinpoint a particular food product's movement or characteristics. In some cases, the objectives of the system will dictate a precise system, while for other objectives a less precise system will suffice. In bulk grain markets, for example, a less precise system of traceability from the elevator back to a handful of farms is usually sufficient because the elevator serves as a key quality control point for the grain supply chain. Elevators clean and sort deliveries by variety and quality, such as protein level. Elevators then blend shipments to achieve a homogeneous quality and to meet sanitation and quality standards. Once blended, only the new grading information is relevant—there is no need to track the grain back to the farm to control for quality problems. Strict tracking and segregation by farm would thwart the ability of elevators to mix shipments for homogeneous product.

The basic component of the traceability system:

There are six important elements of traceability which put together, constitute an integrated agricultural and food supply chain traceability system:

(a) *Product traceability* - which determines the physical location of a product at any stage in the supply chain to facilitate logistics and inventory management, product recall and dissemination of information to consumers and other stakeholders.

(b) *Process traceability* - which ascertains the type and sequence of activities that have affected the product during the growing and postharvest operations (what happened, where, and when). These include interactions between the product and physical/mechanical, chemical, environmental & atmospheric factors which result in the transformation of the raw material into value-added products; and the absence or presence of contaminants.

(c) *Genetic traceability* - which determines the genetic constitution of the product. This includes information on the type and origin (source, supplier) of genetically modified organisms/materials or ingredients as well as information on planting materials (such seeds, stem cuttings, tuber, sperm, embryo) used to create the raw product.

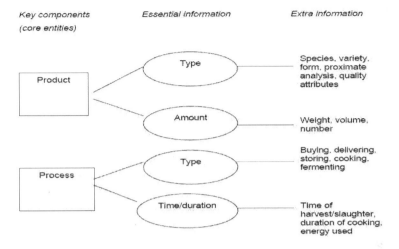

(d) *Inputs traceability* - which determines type and origin (source, supplier) of inputs such as fertilizer, chemical sprays, irrigation water, livestock, feed, and the presence of additives and chemicals used for the preservation and/or transformation of the basic raw food material into processed (reconstituted or new) food products.

(e) *Disease and pest traceability* - which traces the epidemiology of pests, and biotic hazards such as bacteria, viruses and other emerging pathogens that may contaminate food and other ingested biological products derived from agricultural raw materials.

(f) *Measurement traceability* - which relates individual measurement results through an unbroken chain of calibrations to accepted reference standards.

To achieve this, measuring and test equipment and measurement standards are calibrated utilizing a reference standard whose calibration is certified as being traceable to a national or international standard16. The other aspect of measurement traceability relates to the property of the measurements (data and calculations) generated throughout the supply chain and their relationship to the requirements for quality. By focusing on the quality of measurements (rather than on a property of an instrument, it is possible to assure that the measurements are indeed adequate for the intended use. To achieve this, each measured data must specify the environmental, operator, and geospatial and temporal factors, which are not related to the instrument but impact on the quality of the data. In implementing a new traceability system or studying an existing one as part of routine quality management system or in the event of food safety and quality alert, these basic aspects must be addressed in order generate sufficient data to adequately evaluate the type, origin and location of the source of safety concern to enable corrective actions to be taken.

Traceability is an information-based proactive strategy to food quality and safety management. It is a complimentary tool to other quality management programmes such as Hazard Analysis and Critical Control Points (HACCP) systems. A key strength of traceability chain management is that it facilitates the identification and isolation of hazards and implementation of effective corrective actions in the event of an incident. Thus, like point inspection and product testing, traceability by itself cannot introduce safety into the food process or handling process. When considered in isolation of other quality management systems, its traceability is not a sufficient condition to satisfy the safety requirements of the food chain. However, its strength lies in preventing the incidence of food safety hazards, and reducing the enormity and impact of such incidents when they occur by facilitating the identification of product(s) and/or batches affected, specifying what occurred when and where it occurred in the supply chain, and identifying who is responsible.

The benefits of integrating traceability into the overall quality agricultural management system are numerous, ranging from improvements in product quality and safety management, crises management in the event of a safety alert, and strengthening overall agribusiness coordination. With heightening public scrutiny of the food supply chain and agriculture, many national and regional new food quality regulatory directives and laws have been enacted, leaving agriculture and food industries with little option but to implement traceability systems as part of the overall food safety and quality management programme. As agriculture continues to experience declining terms of trade and competition by other more financially lucrative industries, there are good reasons to believe that the concern about traceability will continue in global food trade. The search for cost-effective technological innovations for implementing accurate and reliable traceability systems is therefore an important challenge facing agriculture in the new globalised economy

6. *Food Traceability Frame work:*

A product traceability system, and particularly a food traceability system, is fundamentally based on 4 pillars: product identification, data to trace, product routing, and traceability system and stresses this concept.

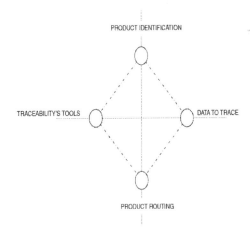

PRODUCT IDENTIFICATION	DATA TO TRACE	PRODUCT ROUTING	TRACEABILITY'S TOOLS
dimensions	number	Production cycle	compatibility vs product
volume	typology	activities	compatibility vs process
weight	degree of detail	lead times	N° of data readings
surface conditions	dynamism	equipments	N° of data writings
shortness	data storage requirements	manual operations	degree of automation
perishability	confidentiality & pubblicity	automatic operations	data accuracy
packaging	checks & alarms	movement systems	data reliability
cost		storage systems	company's knowledge
life cycle length			cost of system
bill of material structure			

The data accuracy and reliability required can guide the selection of the traceability tool. Obviously, cost is a relevant factor and so must also be taken into account. The figure shows a particularly complete framework for product traceability. Even though each product has different characteristics, a traceability system can be derived from general framework for a product by using a process of simplification. For example, not all products have the perishability problem.

Food products are very critical, and their traceability system must be particularly complete: in this case the entire framework in figure represents an efficient system and a very significant reference

7.Traceability System:

Food traceability systems are record keeping procedures, or tracing systems, that record the path of a food product or an ingredient in a food product from its initial supplier through all processing stages until it reaches the end consumer. A traceability system allows the food industry to: promptly locate and remove unsafe products in case of a recall protect brand reputation — keeping precise records allows companies to quickly identify and recall only unsafe products, reducing the scope of a recall, demonstrating good corporate citizenship and a high level of concern for public health, therefore limiting negative media exposure and perhaps even turning it positive minimize the size of a recall and reduce the cost incurred in recovering or disposing of products in the market place diagnose problems in production and determine liability where relevant — traceability can help resolve process problems and determine third-party responsibility if records show that an ingredient supplier or co-packer was the source of the recalled ingredient. Although the manufacturer of the final product is still responsible for the recall, complete records tracing ingredients to their sources may allow seeking indemnification from responsible third parties

Types of Traceability Systems:

Traceability systems can be manual or computer based. Small companies manufacturing limited numbers of products with simple formulas, short shelf-lives and fewer customers may find paper-based, manual systems adequate. Large companies may find computerized systems more reliable and efficient. Computerized systems can help:

- ✓ speed data and product handling
- ✓ reduce errors
- ✓ reduce paper waste
- ✓ track product movements precisely

Alphanumeric Notes

Alphanumerical codes are a sequence of numbers and letters of various sizes placed on labels, which in turn are placed on product or on its packaging. Clearly, in the food sector the latter of these two practices is the most suitable. The design phase of this system is very simple and economic, but its management requires significant human resources (and so costs) because code writing and code reading are not automatic. Furthermore, performance is not particularly good: there are many problem associated with the large amount of managed manually data. The risk of data integrity corruption is very high. No standards are defined for alphanumerical codes, and they are generally ''owners'' codes, so there is a unique and not general tie between the different actors (raw material suppliers, manufacurers and distributors) in the supply chain analyzed. The European Article Numbering (EAN) association has made some effort towards standardization by introducing several codes: the best known is the EAN/UCC Global Location Numbers (GLN) in the EAN/UCC-13 version. Today, alphanumerical codes are not frequently used because bar codes offer several significant advantages.

Bar Codes

In effect, the introduction of bar coding has modified handling of all materials along the supply chain and moreover particularly affects the traceability question. The automation, the high speed, the great precision (it is a practically error free system) guaranteed by a bar code structure permits simpler, more economical, and exact traceability systems. At the time of writing more and more industries, especially in the retail sector, use bar codes as a principal means of identifying items. Various applications of industrial product traceability (non food) that still work well are based on this technology. With regard to food, traceability is less advanced and there are only a few spot applications such as the case of Anecoop in Spain (Vilanova, 2001).

In a bar code system, each time items are moved from one point to another, their bar code labels must be so positioned that they can be detected and identified by the reader. This characteristic, often called line of sight positioning requirement, requires human intervention (thus time and effort) for the scanning process and so there is room for error and inefficiency. Besides the physical support provided by a label, it usually has generous dimensions and is easily damaged ("optical" damage is sufficient). As a result, bar codes are less attractive to the food sector, and their application is consequently limited.

Radio Frequency Identification Devices (RFID)

In addition to bar code technology, there is radio frequency identification – RFID system. RFID is an identification tool using wireless microchips to create tags that do not need physical contact or particular alignment with the reader. The reading phase is very fast and fully automated. RFID tags are very small (a few millimeters reading distance) and they have no compatibility problem with foods. The TAG is an isolated system, their materials are aseptic and food compatible. The link between TAG and product is very easy: for solid goods gluing system is very effective (glues are absolutely neutrals); for liquids TAG is usually connected to storage or package system. The radio wave used for communication between TAGs and traceability database use very little power, so electromagnetic interaction is practically non-existent. Latest technology at the time of writing has a good data transfer rate, even when a great deal of electromagnetic interference affects the ferromagnetic field.

✓ A reduction in labor cost. RFID simplifies handling and storage processes, particularly as no manual scanning and checking operations are required.

✓ An acceleration of physical flows. As an RFID reader can scan numerous tags at the same time, identificationis very simple and rapid.

✓ A reduction in profit losses. The University of Florida concluded that nearly 2% of total sales in United States is lost each year due to "shrinkage" – employee \and custom er theft, vendor fraud, and administrative error.

✓ More efficient control of supply chain in terms of improving control of the stock situation, and production monitoring.

✓ Improved knowledge of customer behavior This knowledge has great importance especially for new products or items in a promotion for which it is not only important to check whether or not they are selling, but also to know whether or not they are being taken away but not bought by consumers.

With specific reference to the food sector, RFID is a very promising system because it also results in:

✓ Improved management of perishable items. The continuous monitoring of item routing reduces waste and improves customer service levels.

✓ Improved tracking and tracing of quality problems. In using individual product codes, RFID systems are providing means to identify and find only defective product, and so help react to any quality problem.

✓ Improved management of product recalls. The ability to trace product routings can secure efficient recall procedure and help producers and distributors to minimize damage.

However, some RFID properties limit traceability systems. The main problem relates to tag cost. A tag costs between 0.5 € and 20 €, whereas the bar code is a low-cost system, so RFID is more expensive and can represent a significant handicap for product final price, particularly for low-price products (e.g. fruit, vegetables, pasta, milk, etc.). In addition to this problem, there are some operational questions producing minor difficulties: lack of standardized RFID protocols (the best and most used are: ISO 13 MHz and EAN/UCC GTAG) and scanning problems due to interference under particular electromagnetic conditions. In conclusion, at the time of writing the best technical instruments to use for a product traceability system are bar codes and RFID systems. In particular, RFID presents very favorable properties for the food sector, but the tag cost remains a problem.

8. Traceability Process Steps:

Establishing traceability is a three step process: They are:

Step 1: Unique Identification in the value Chain

Step 2: Data Capture & Recording

Step 3: Establishing Links

Step 1: Unique Identification in the value Chain

Any object that goes into production of a consumable finished product needs to be identified & tracked at all points in the supply chain. The depiction adapted from the industry standard SCOR (Supply-Chain Operations Reference) model for supply-chain management explains the interaction points between various entities starting from supplier's supplier right up to the final consumer. Between two levels, entities perform standard operations of sourcing, making and delivering.

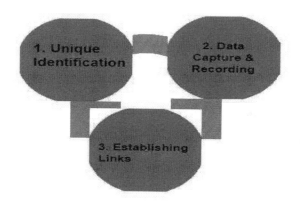

Traceability Process Step

One-up & One-down traceability is one of the typical regulatory requirements in most of the countries. Every entity in the chain should keep track of the upstream source of their materials and the consumption point down-stream. To comply with the one-up traceability, a manufacturer who uses sugar as one of the ingredients should keep track of the supplier from whom he sourced sugar & the production batch that consumed sugar. To comply with the one-down traceability requirement, he needs to keep track of the supplies of finished goods that were sent to distributor warehouse. Extending the same to all partners in the chain, we achieve full value traceability. The foundation of traceability lies in identifying every object uniquely across the chain. As there are a lot of players involved in getting the products to the final consumer, and each of them have their own priorities.

Step 2: Data Capture & Recording

The overall scope of traceability should be to capture enough data to have a full genealogy on parts and processes of the organization. The amount of data that the organizational systems can rake in will define the granularity to which they can get down to nail a casualty's root cause.

At a high level, the flow in a food supply chain is Supplier -> Manufacturer -> Distribution centre -> Wholesaler/distributor -> Retailer -> Consumer. Major entities of a food supply chain from a data capturing (using RFID tags) perspective are Manufacturer, Distributor and Retailer. Broadly, the roles of each of the entities are:

- **Manufacturer:** The manufacturer creates an initial catalog based on his requirements for tagging the raw material/ingredient. *Catalog is a secure file that stores data about each move a product makes through the food supply chain.* Catalogs can help reduce counterfeiting of food and other products. The manufacturer then tags the raw material /ingredients procured from the supplier.
- **Distributor:** The distributor confirms on the shipments received using the catalog sent along with an Advance Shipment Notice (ASN) by the manufacturer.
- **Retailer:** Retailer confirms the shipment received using the catalog sent by the distributor. In today's scenario, there are a good number of methods available

to capture data. Starting from conventional methods of capturing data manually to automated methods to pull/push models of data capturing exists. The various methods of capturing data include:

Manual tracking

- HMI /SCADA
- Electronic Identification
- Bar Codes
- RFID

Step 3: Establishing Links

The technical impact of traceability becomes obvious as the system links all the information that is captured throughout the supply chain. The tagging or naming could be a typical system generated number /sequence. All the tagging numbers will hold one or more key values that can be directly mapped to the numbers generated in the previous link in the chain, providing us with a mechanism to trace upstream or downstream in the chain. A deeper system will enable the establishment of links among more agents further up or down a supply chain.

Design of traceability system

A traceability system should be designed within the context of a broader management system to the organization. The choice of a traceability system should result from technical feasibility and the economic acceptability. The traceability system must be verifiable. The documentation and procedures are to be established by considering the regulatory and policy requirements. During designing, the flow of the materials should be considered and the information required for the traceability system must be identified. The coordination among feed and food chain should be developed.

Steps for designing traceability

✓ Identify the suppliers and customers and determine the position in the food chain.

✓ Determine and document the flow of materials

✓ Establish procedures by documenting the flow of materials and related information, including document retension and verification.

✓ Establish the requirements of documentation to achieve the objectives of the traceability system.

✓ Establish coordination in the feed and food chain.

Use of traceability in Product recall:

A traceability system serves many purposes, the important ones being product recall. The processes should be designed in such a manner to assist manufacturers, suppliers and importers to quickly and effectively track and trace the details of the product and to communicate for the initiation of the product withdrawals and product recalls. It also assists retailers and wholesalers to effectively contact regulatory agencies. Internet based services for supply chain information exchange from source to shelf. The system collects stores and reports forward and backward traces and between supply chain.

Mock traceability:

The mock test is very important to strengthen the preparedness of the organization in the time of real crisis. The better the preparedness lesser the extent of the damage. This is to check the time taken for the internal and external traceability and the efficiency as well. The planning of the mock traceability process includes defining the product and / or the relevant ingredients, defining the lot, documenting the flow of materials, including the media for documentation , managing the data and retrieving information for communication. At the time of mock traceability, the process should include defining the product, identifying the lot, documenting the flow of all materials at each steps, retrieving the process and quality data, packaging and storage, the process details, the environment, the situation, retrieving the other information at the time manufacturing, any abnormality tracing reports on the flow of the raw material from suppliers and finished goods to the consumers and proof of implementation of a management system regime. The efficiency of the mock traceability is measured based on the extent of information we can trace and time taken to get these information.

9.Internal traceability in food manufacturing

Historically process and quality control systems were operated separately in food manufacturing. However, integration is now common and indeed the whole manufacturing process from recruitment of staff, through manufacturing to marketing of products is now increasingly handled as a single integrated process. The development of traceability systems in part reflects this change. Paper–based systems are still widely used for traceability systems in both large and small companies, and even within systems operating across the whole food chain. Manual entry of data to databases through the production process or at regular intervals from hand-written records is also common. Often these databases are linked to other current software packages *e.g.* accounts management systems or stock handling and reconciliation systems.

Enterprise Resource Planning Systems (ERP) and bigger e-collaboration systems are large business management systems which operate at all levels within a company: inventory management, operation planning, sales planning, maintenance, document control, quality control, human resource management, salaries and much more. Investment in such systems has been a major commitment within all sectors of manufacturing in the last decade. Different suppliers operate to deliver ERP systems to customers of various sizes, so that company size is not a limiting factor for the introduction of an ERP system. Where installed the ERP system is the main driver throughout the factory for ordering of goods, manufacturing planning and financial systems. The system is accessed on networked terminals throughout the company. Different users are given varied levels of access according to their password. Data entry can be cross referenced using internal checks so that errors are difficult, while not absolutely impossible. At any point following receipt of goods their position in the system is indicated by ERP system. Traceability is inherent within the system and necessary for much of the functionality *e.g.* measuring the efficiency of process management. However, it is rarely one of the main goals. The ERP system provides a one-stop shop for information, which can be very quick in case of emergency. It may take only minutes to identify all related batches through their ingredients etc. The development of ERP systems to deliver both forwards and backwards traceability as a key outcome is on-going.

Companies supplying industrial automation equipment are also increasingly demonstrating how it can be used as part of an integrated system of products, technologies and support services to deliver traceability. Data collection throughout the production process is becoming more common and more automated. Machine readable identification systems can allow in process traceability to be demonstrated. Programmable logic controllers can be used to establish communication between separate process elements throughout the plant. This information can be managed and interpreted to deliver critical data relating to the process control points, as well as data which can be used to derive yields and process efficiency information. Specific food safety/quality management systems have been developed, these include laboratory information systems. Some systems act as approval systems for raw materials. Where details of suppliers and emergency contacts can be stored on this database, some backwards traceability can be delivered. Other systems are used to handle recipe formulation and raw material information for complex products and compile information on individual product specifications. This is coupled with information obtained through the manufacturing process to create the product reports required for each customer. IT based systems have been introduced in this area to give greater accuracy and control of information while reducing transcription errors and reducing the time taken by 90% compared to paper-based systems. It is clear that different systems relating to traceability in the manufacturing process have been introduced at different times for different reasons. This may create data islands within the company, where data is held on separate computer systems or in different filing cabinets in separate offices. For most efficient operation of traceability systems, rapid access to all data from the production process from raw materials purchase to product distribution presented in a seamless and cross referenced manner is essential. There are now IT–based systems, which have been developed with the specific aim of fully integrating information from other management systems (ERP, specialist process control and laboratory information management systems) at a range of scales. These integrating systems seek to deliver control in the management of quality throughout the manufacturing process by linking up and integrating data collection and recording systems throughout the plant/operation giving real-time validation to ensure compliance with quality procedures and guaranteeing full backward and forward traceability. Such systems may give greater control throughout the manufacturing process. Goods can be held until conformance tests are passed and products can be directly supplied to the different specifications of a range of customers.

Many of the difficulties in the implementation of traceability systems within manufacturing systems are related to determining the traceable product unit size (what size should the batch be?), which is determined by the details of the particular process management. Some of the key areas of difficulty identified were:

Continuous and batch processing and the transfers between such processes within the manufacturing system. The handling of **bulk products** (sugar, salt, glucose syrup, flour).Even where goods are delivered with clear batch identification in a tanker, they may be emptied into a single silo and mixed with earlier deliveries, so onward traceability may not be maintained. Silos also have dead zones in filling and emptying, which can cause the blending of successive batches. Changes in handling practices can increase traceability and telemetering can be used to carry forward any batch identification system provided a delivery into the process.

Rework of any component of the recipe expands the traceable unit, so that whatever other traceability systems are in place, the size of traceable unit is actually all the product produced between two breaks in the rework cycle. Just because any rework is traceable, does not mean that the contaminant is not dispersed through several days/months/years or production. Any rework cycle can have significant implications for product recall

Water used in food processing and manufacture. Are records of date and time of manufacture good enough?. Really detailed application of traceability systems is always possible, but such an

implementation in any business could be very high cost and of little benefit for consumer safety, if the next step in the chain does not maintain or improve on the size of the traceable unit. Risk assessment is the key step in the design of a traceability system which enables the management of risk, dependent on the raw material purchased and the final product that is produced. Tracing absolutely everything to the nth degree may be an unnecessary safeguard. Companies currently investing in integrated quality management systems, which integrate information and deliver rapid, validated traceability are more likely to be :

- ✓ seeking to use product quality as a differentiating criteria in the market place
- ✓ seeking to manage/reduce risks to shareholders and have recognised quality as a

   major factor
- ✓ seeking to reduce the impact of product recalls on business by reducing recall

   widths and minimising the need for recalls from customers
- ✓ working in complex industries where the provisional of batch specific quality information is critical to ensuring a market (shell fish based on safe harvest zones defined by regular Local Authority analysis)
- ✓ working to achieve/maintain an ISO9000 standard within the company
- ✓ seeking to reduce the impact of preparation for and audit visits from suppliers by always being ready
- ✓ seeking benefits in improving yields and performance in the process by reducing waste and constantly improving process quality

   "They won't buy the system unless they are really committed to quality"

10. Linking identification and information

Product withdrawal and recall systems only require traceability in part of the chain from the production step to the consumer. However, if the problem stems from the raw material, traceability back to the supplier improves the possibility of correcting fault, avoiding recurrence and/or placing the responsibility (and liability) there. So traceability usually functions both forwards and backwards through the chain. There is currently a large gap between the strict quality and traceability standards applied to pharmaceuticals (e.g Food and Drug Administration requirements) and those applied in the food industry. However, the industry believes that in many sectors of food manufacturing this gap will close, through the application of technology in process and quality control from the pharmaceutical industry. In much of the food industry much work has recently been done to develop customer/supplier relationships and create assured chains of supply. However, given that unforeseen events can always happen, traceability systems are still in place even where strong customer/supplier relationships exist.

11.Within a business – process traceability

There is a range of systems for traceability in use across the industry. However, their basic characteristics are the same, whether the systems are IT enabled or paper based. In large manufacturing businesses, full business systems underpinned by large computers or networks are becoming more common. However paper-based systems may be used to link the product identification and associated information even in the largest companies. *"Most important is that traceability is built into working practices and culture"*. Figure 3 illustrates the links between identification and information through a manufacturing process. On receipt goods are booked in, their specifications checked and they are given a lot number. From the store, raw materials will be booked out to the factory as they are needed and their use is linked to manufacture (by date and time of use, works order, recipe etc.). The final product is given a traceability/lot code (date code, time of manufacture, lot number), which allows production to be grouped and gives a link back to the raw materials used, time of the production run etc. For dispatch, products may be grouped onto pallets, or into boxes, which may be given separate identification codes for use in dispatch and subsequent handling in the supply chain. While in overview traceability systems in food manufacturing are similar, their detailed implementation is different and closely related to process management. In continuous processes, manufacturing windows can be defined by a regular sampling pattern (for quality analysis) at some point in the process, where the time taken for the product to pass through the process identifies these windows further down the line. Date/time markers may also be placed regularly into a production line and their appearance at later stages in the process can be recorded and create manufacturing windows. Even for continuous processes, batches are usually created for despatch – *e.g.* tanker, box, pallet of bags. Batch codes created at this point often record the date and time of packing and may also indicate the packing line or silo from which the product was loaded.

Figure 3 Schematic diagram showing linked information within a process traceability system from raw material receipt (at the bottom) to the finished product awaiting despatch (at the top).

The larger the window used for batching the simpler the system is to operate, but the larger the amount of product, which may require being withdrawn. The size of the traceability windows for recall may be as wide as a day's or as narrow as 15 minutes production. The size of this window is a business decision based on what is practical in relation to process management and also on wider considerations relating to business risk management.

The detailed and complex information in traceability systems lends itself to recording, compiling, transfer and interrogation by electronic means and there is an increasing move towards making traceability systems IT enabled. IT enabled systems may also use data-scanning by incorporating bar codes or RFID identification systems which can be read automatically for raw materials or during the process. These inputs can speed up data recording for traceability and reduce errors in data entry. Some components of a fully automated system may be expensive; most systems are designed so that they can be implemented in a staged way. The cost of training may also be high in initial implementation phases, but it is critical that the system is understood and fits into the working pattern of the staff.

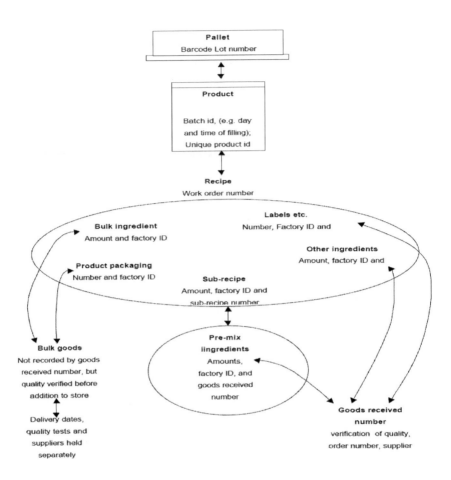

12. Traceability in Mineral water plant:

Mineral water is defined as water which comes from a natural source and has at least 250 ppm total dissolved salts. Mineral water, which is well-represented amongst bottled waters, becoming ever more popular. According to the Natural Resources Defence Council (NRDC, 2007) 35 percent of people preferred to drink bottled water because they are worried about the safety of tap water. Bottled water is steadily increasing in popularity despite the fact that questions are often raised about the origin of the mineral water and regulation of the mineral water industry.

In recent years there has been increased focus on traceability in food supply chains. Brand image is important in the food industry, and food business organizations (FBOs) endeavor to establish a good relationship with their customers in order to distinguish themselves from others on the market. An important part of building a brand is informing customers about product qualities that they might not otherwise be able to detect, so called "credence goods or quality goods". These and many other contributing factors have brought traceability to the attention of the mineral water industry.

For example a mineral plant produces 30,000,000 liters of mineral water per year. The products were glass or plastic bottles of still or sparkling mineral water in various sizes. The production capacity for plastic bottles was 18,000 liters of water per day. The customers were mostly local retailers and restaurants. Weaknesses and areas where traceability could be improved at mineral plant are analyzed and recommendations for changes to existing routines and practices to improve traceability. Traceability is not the product and process information itself, but a tool that makes it possible to find this information again at a later date.

Two types of traceability are Internal traceability is the ability to trace the product and process information internally in a company. Chain traceability is the ability to trace the product and process information through the links in a supply chain, in other words the product and process information a company receives from their suppliers and transmits to their customers. here are three types of traceable units: batch, trade unit and logistic unit. A batch is defined as a quantity going through the same processes. In this plant the batch is synonymous with lot. A trade unit is defined by Global Solutions One (GS1 previously EAN.UCC)

(2007) as: "Any item upon which there is a need to retrieve predefined information, and that may be priced, ordered, or invoiced at any point in the supply chain". A trade unit is a unit which is sent from one company to the next company in a supply chain. Examples of such units are a box, a bottle or a six pack of bottles.

A logistic unit is defined by GS1 (2007) as an item of any composition established for transport and/or storage that needs to be managed through the supply chain. The logistic unit is a type of trade unit, and it designates the grouping that a business creates before transportation or storage. The classic logistic unit is a pallet, but it may also be a container, a boat load, or similar. Directly linked means that the information about different processes or resources linked to a product is recorded in one place, for example on one sheet of paper and/or software program. Systematic information loss occurs when information about a product or process is not directly linked to a product and recorded systematically. A place where systematic information loss occurs is often also a critical traceability point (CTP). A standardized globally unique identification (ID) can be used to identify trade units and logistic units. Local or non-standard ID is identifiers which are only unique and meaningful within a company.

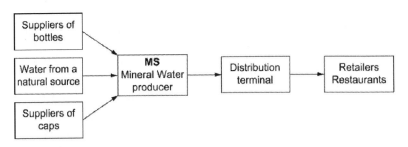

Overview of the material flow in the mineral water plant

MS received water from a natural source. Each day the water was pumped into one of two tanks. Each of the two tanks used had a capacity of 500,000 liters of water. Twice a year these tanks were emptied and cleaned. MS recorded each time a tank was filled with water. During the production of mineral water MS recorded production date and tank number. This number was not directly linked to the bottle of mineral water. Production date could be found from the "best before" date on the bottles. The bottles were delivered by truck from multiple suppliers. MS received 30 pallets in each delivery. Each pallet consisted of one box of bottles, identified by a box number. Some bottles were labeled with the supplier's name and batch number. When bottles were delivered MS recorded the supplier's name and the box number. Box numbers were recorded manually when the bottles were processed.

The number of the box that bottles were taken from was recorded and indirectly linked to the "best before" date found on the bottle. The screw caps were delivered by truck from multiple suppliers. The boxes of screw caps were identified with box numbers. 25 pallets arrived in each delivery and an ID was given to all the screw caps in each delivery. MS did not record or link any information about the suppliers to this new internal ID.

Bottles of mineral water were marked with a "best before" date. The production on one day often ended with a half full pallet that was completed the next day. Thus a pallet could consist of bottles with different "best before" dates. A pallet with bottles of mineral water was identified with a serial shipping container code (SSCC), GS1 14 (GS1, 2007) and lot number and bar code. Before delivery, truck drivers recorded electronically the pallet number and the destination of the delivery. MS could identify the customers through the date, but not with 100 percent accuracy.

To trace the water back through production is possible because MS recorded production date and tank number, which was linked to the "best before" dates on each bottle. The water could be tracked back to a specific tank on a specific day. If necessary the bottles could be identified and recalled directly. This recall would use the "best before" date. This means that the only relevant and possible good traceable unit for water was the water that is in a tank on a specific day. The link to the tank number was always available through the production date, as each individual bottle was always marked with a "best before" date.

It is currently possible to trace mineral water back and track water forward through MS with the existing system. MS fulfils the current EU requirements for traceability. With respect to traceability, production was simple with separate batches for mineral water. Practical changes to the size of units which are used for traceability, for example using pallets of screw caps rather than entire deliveries would enable more precise recalls to take place if required.

13. Traceability in Chocolate industry

In the chocolate industry, raw material batches from different suppliers may be mixed. If a food safety problem comes from a certain raw material batch, all finished products containing raw materials from that batch have to be identified and recalled. Thus, the magnitude of a recall directly depends on batch dispersion in production and distribution. Recently, researchers found that the best solution to reduce batch dispersion is to reduce processing batch size and batch mixing. However, it was also found that reducing batch size leads to losses in production efficiency, due to increased production setup times, setup costs, cleaning efforts, etc. Traceability system focused on the supply chain of chocolate and the production system.

Supply chain of chocolate manufacturing

The supply chain of chocolate starts with cocoa farming. Cocoa farmers grow, harvest, ferment and dry the cocoa beans. After packing the dried beans in bags the cocoa farmers deliver the cocoa beans to local buying stations, which combine the bags of several cocoa farmers and deliver them to a cocoa exporter. Cocoa exporters combine the bags of several local buying stations into batches. The cocoa exporters organize the shipping to the chocolate manufacturer. The different process adopted in chocolate manufacturing includes i) different production strategies ii) different traceability system iii) different product recalls

Different Production strategies

The chocolate production system for two different production strategies, one based on production efficiency (PS1) and one based on reduced batch dispersion (PS2). In PS1 the maximum processing batch size is always used so that the equipment in the production stage is always used at full capacity. Since the size of the cocoa bean batches delivered to the chocolate manufacturer is not necessarily a multiple of the processing batch size, some cocoa beans are mixed with the next batch of cocoa beans. This results in having some batches of finished product produced from two different batches of raw materials. Instead, PS2 focuses on reducing batch dispersion, where the chocolate manufacturer avoids mixing the different batches of cocoa beans. Here, some processing batches might be smaller in size. As batch processes are involved, this results in some partially unutilized processes in the chocolate production line, with a corresponding reduction in production efficiency. On the other hand, if a safety crisis occurs to a batch of raw materials, a PS2 production strategy would lead to smaller recall sizes compared to PS1. A graphical illustration of both PS1 and PS2 can be seen in Fig. 2. The production efficiency is measured by the number of processing batches because:

i) The number of processing batches equals the number of times a roasting process is performed and the duration of the roasting process depends on the roasting grade desired, not on the amount of nibs processed into the equipment.

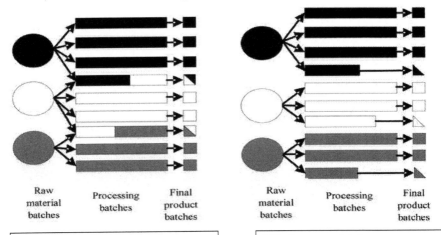

PS1 –Raw material batch sizes are mixed in order to utilize maximum processing batch capacity

PS2 –Raw material sizes are never mixed

The processing batches mean less time needed for roasting, with a constant number of equipments; or less equipments needed, with a constant processing time required. Thus, less processing batches lead to a higher efficiency. Smaller batch sizes (also meaning more batches when processing a constant raw material amount) lead to an increase in production setup times and costs, resulting in losses of production efficiency.

Different traceability systems

It includes a basic traceability system (TS0), and two improved traceability systems (TS+ and TS++). TS0 fulfils the European law regarding traceability, thus the actors involved in the supply chain and located within the European borders follow the "one step back-one step forward approach" required by law. That is, the finished chocolate product is traceable from the supermarket, to the chocolate manufacturer, to the cocoa exporter. As the cocoa exporter is assumed to be located outside country it is not possible to trace the cocoa beans further in the supply chain. TS+ is an extension of TS0, where the local buying stations, when buying the cocoa beans from the cocoa farmers, mark all cocoa bags with a unique code and the buying date. Also, when the cocoa exporter buys the cocoa and mixes bags from different local buying stations, the original bags remain. This means the cocoa is delivered to the chocolate manufacturer in the original bags with the code of the local buying station. These codes are then registered so that the finished chocolate can be traced up to the local buying station.

TS++ extends TS+ with the addition that the cocoa farmers, when packing the cocoa beans, already mark all bags with unique codes and date. In this case the finished chocolate is traceable up to the individual cocoa farmer. Alternatively, the local buying station could mark the bags at arrival, with the information of the farmer delivering the beans.

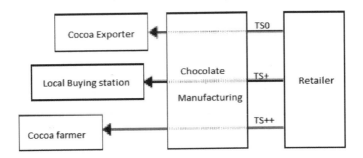

Different product recalls

The two possible food crises and corresponding recalls (R1 and R2) that could occur in the supply chain. R1 simulates the product recall in case of a contamination of the cocoa beans, which could be a chemical contamination while farming, fermenting or drying. In this case all chocolate bars produced with cocoa beans from a certain cocoa farmer need to be recalled. R2 simulates the product recall in case of a contamination of a processing batch, which could be caused by a problem in a roasting process. In this case all chocolate bars produced in a certain roasting process need to be recalled. The simulation models allow to run single and multiple simulations. Due to the importance of the roasting process it is also possible to run single or multiple simulations automatically for different processing batch sizes. For this paper, we simulated the food scares for a range of processing batch sizes between 1,600 kg and 5,000 kg
(every multiple of 200 kg). Each of the sizes is then run multiple times, while information such as number of runs, processing batch size, recall size and number of processing batches (which reflects the production efficiency) is registered, and average results can be determined.

14. Examples of Food traceability tools and Labels:

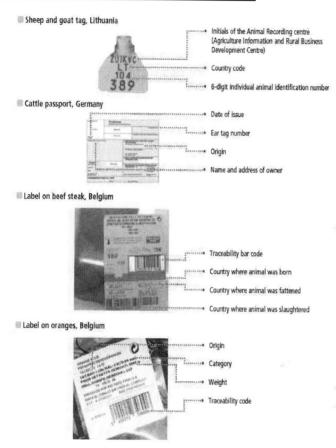

15. Conclusion:

- Quantitative Results on real or hypothetical improvements thats helps food industry in decision to improve their traceability system.
- Traceability system is a highly valuable tool from which it is well designed and well integrated with the production operations

- Food producers benefit in terms of safety issues as well as in the daily operational activities.

16. Reference:

1. Moe, T. 1998. Perspectives on traceability in food manufacture. ***Trends in Food science and technology.*** pp. 211-214.
2. Pradip Narayan Das. 2011. Traceability –It's importance in Food Industry. ***Beverages and Food world.*** pp. 27-28.
3. Rolando, S and A. Renzo. 2011. Testing improvements in the chocolate traceability systems: Impact on product recalls and production efficiency: ***Food Control.*** 23: 221-226.
4. Vinny, S and Amarjeet Kaur, S.S. Thind and K. S. Minhas. 2011. Traceability in Food supply chain from "Farm to Fork". ***Indian Food Industry.*** 30: 5-6.
5. Kine, M.K., O. Petter and Kathryn, M. 2010. Implementing traceability: practical challenges at a mineral water bottling plant. ***British Food Journal.*** 112(2): 187-197.
6. Regattieri.A , M. Gamberi,and R.Manzini. 2007. Traceability of food products: General framework and experimental evidence. ***Journal of Food Engineering.*** 81: 347- 356.
7. Traceability in Food and beverage industry.
8. Handbook for introduction of food traceability systems.

i want morebooks!

Buy your books fast and straightforward online - at one of world's fastest growing online book stores! Environmentally sound due to Print-on-Demand technologies.

Buy your books online at
www.get-morebooks.com

Kaufen Sie Ihre Bücher schnell und unkompliziert online – auf einer der am schnellsten wachsenden Buchhandelsplattformen weltweit! Dank Print-On-Demand umwelt- und ressourcenschonend produziert.

Bücher schneller online kaufen
www.morebooks.de

VDM Verlagsservicegesellschaft mbH
Heinrich-Böcking-Str. 6-8   Telefon: +49 681 3720 174   info@vdm-vsg.de
D - 66121 Saarbrücken   Telefax: +49 681 3720 1749   www.vdm-vsg.de

Printed in Great Britain
by Amazon

85800126R00036